Notes:

Steel Sailboat Masts

Design and Construction

D. L. Schaffer

The methods detailed in this book reflect the authors' own experience. The author assumes no responsibility for the use or misuse of the information contained herein.

Copyright   2022

No part of this book may be reproduced, stored in a retrieval system, or transmitted, in any form or by any means, electronic, mechanical, photocopying, recording, or otherwise, without the written permission of the publisher.

Contents:

Introduction: Page 6

Steel Mast Design: Page 11

Steel Mast Components: Page 16
Mast Column: Page 18
Mast Head: Page 22
Spreaders: Page 29
Mast-Boom Gooseneck: Page 35
Thru-Deck: Page 38
Mast Base: Page 43
Standing Rigging: Page 48

Fabrication of the Mast Column: Page 51

Bezier Designs: Page 56
Bezier Books: Page 61

Introduction

Way Back, when I was young and building my first steel boat, I chose to also fabricated the steel mast. I thought that the money spent purchasing a commercially built mast could be better spent on products that I could not fabricate myself.

I was a skilled metal fabricator, but at time my engineering knowledge was limited, consequently I did a little research. In one well-known book on Yacht Design, I read where Nathanael Herreshoff experimented with steel masts.

In another I learned how 'Euler's Slender Column' formula was used to size sailboat masts. The interesting part was that Euler's included an option to design in steel, pivoting around the 'Module of Elasticity' and 'Moment of Inertia'.

I therefore, decided that I would fabricate a steel mast for my first steel build using the overall section dimensions given by the Designer, with a twist. The mast would be eight sided.

That mast, shown in the below pictures, stood on the 'Notion' for the twenty some years that I sailed her. At over twenty years old her new owner successfully crossed the Atlantic, with that eight-sided steel mast that I built so many years ago.

If you are a 'One of' builder, look at the time that it takes to build a steel over the time that you have to work a job to purchase a manufacture mast and all the hardware.

- Either you are going to work a job to earn money to buy a commercially fabricated mast system.
- Or, you could spend that time building a Steel Mast.
- In either event you are going to spend time. It is just a matter of where you chose to spend your time.

If you are a steel 'Commercial Boat Builder' think back to the time when sailboat masts were made of wood, and were just another part of the build. ***I am thinking, Steel boats – Steel masts, just another part of the build.***

Notes

Steel Mast Design

The calculations and construction used throughout are based on the hydrostatics of a 34-foot Double Hard Chine design shown below. It is the hull that I converted to a True Round design in my book **'Converting Hard Chine Sailboat to True Round'**.

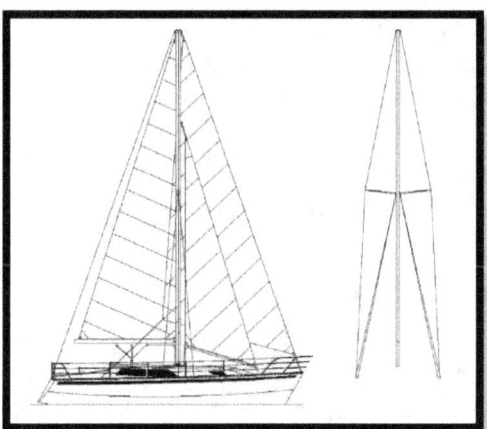

Compressive Load Calculations:

There are two methods used to calculate the compression on a mast. An Empirical method, which is based on the Righting moment at 30 degrees and the beam at the chainplates and Euler's, which is based on the force triangle of the unsupported panels. This writing I will use Euler's because it accommodate masts designed in aluminum, wood, and steel, as the Empirical method restricts itself to aluminum and wood.

Several mast section wills be calculated for this boat:

- The commercially available Aluminum Mast Section
- Oval Wooden constructed Mast Section – Standard Wall
- Rectangular Wooden constructed Mast Section – Standard Wall
- Fabricated Steel Rectangular Mast Section – 0.062" Wall
- Fabricated Steel Rectangular Mast Section – 0.072" Wall
- Fabricated Steel Oval Mast Section – 0.072" Wall

The Physical Dimensions compared are:

- Mast Material
- Mast Section – Rectangular and Oval
- Mast Wall Thickness
- Outside Longitudinal Dimension

- Outside Transverse Dimension
- Weight per foot
- Total Weight

The Engineering Calculations compared are:

- Longitudinal Inertia
- Longitudinal Safety Factor
- Transverse Inertia
- Transverse Safety Factor
- Righting Moment at 30-degrees
- Angle of Negative Stability

Proprieties of Materials:

The **'Moment of Inertia'** is an indication of the stiffness due to the shape of the material. The shape having a higher moment of inertial will deflect less linearly when placed under a load. To increase the 'Moment of Inertial' increases the length, width, and wall thickness of the mast in Section.

The **'Module of Elasticity'** is 29,000,000 psi for steel, 10,000,000 psi for aluminum and 1,300,000 psi for wood. This means that steel is three times the strength of aluminum and twenty-two that of wood.

Taking into account the above, and the weight of aluminum being one-third the weight of steel, and steel being three times as strong as aluminum, I concluded that the thickness of a steel mast would be one-third the thickness of an aluminum mast. Therefore, the weight of a steel mast would be comparable with the weight of an aluminum mast.

The chart below compares the difference between Steel, Aluminum, and wood as a mast material using 'Euler's Slender Column' formula.

MAST MATERIAL	MAST SECTION	WALL THICKNESS	LONG. DIM'S	TRANS DIM'S	WEIGHT FOOT	WEIGHT MAST	LONG. INERTIA	TRANS inertia	LONG. SAFETY FACTOR	TRANS SAFETY FACTOR	NEGATIVE RA-30
ALUMINUM	OVAL	.187	9.000	6.000	5.04	219.730	37.450	20.020	3.230	3.260	126 31,759
WOOD	OVAL	20%	9.750	7.125	8.980	304.380	282.150	150.670	3.170	3.190	123 30,537
WOOD	RECT	20%	8.625	8.250	8.900	300.720	290.88	152.730	3.260	3.230	122 30,620
STEEL	OVAL	.074	8.625	5.500	5.418	265.030	13.190	8.840	3.300	3.140	125 31,155
STEEL	RECT	.074	7.125	4.750	5.780	281.130	12.927	8.980	3.240	3.290	124 30,953
STEEL	RECT	.062	7.500	5.000	5.110	251.940	12.720	8.680	3.190	3.240	125 31,358

Reviewing the Chart:

Sailboat mast are designed to be three times stronger than they need to be. The chart shows that all the masts reflect this standard as seen in the Longitudinal and Transverse Safety Factor column.

With aluminum being the lightest mast material, what effect does the increased weight of a steel mast have on the stability of the boat. Charted are the 'Angle of Vanishing Stability' and the 'Righting Moment at Thirty-degrees.

The 'Angle of Vanishing Stability' is the angle of heel at which the hull it will not 'Right' itself. Note: To be considered an Offshore Cruiser the minimum angle of heel is 120-degrees.

An aluminum mast, due to its lighter weight, has the greatest 'Angle of Vanishing Stability' at 126-degrees. Wood being the heaviest mast material decreases the 'Angle of Vanishing Stability' to 123- degrees. Steel depending of the shape and thickness, has a 'Angle of Vanishing Stability' between 124-degress and 125-degrees.

The **'Righting Moment in Pounds'** is taken at heel angle of 30-degrees. This is the maximum angle of heel that would be considered normal sailing conditions.

The aluminum mast, being the lightest, has 31,769 pounds of Righting force. If the boat were equipped with wooden mast the righting force would be approximately 30,620 pounds. If equipped with a steel mast the Righting force would be approximately 31,200. It can be seen that the choice of mast material has little effect on the stability of the boat.

Notes

Steel Mast Components

The component hardware attached to a mast are: The Mast Head, Spreaders, Boom Gooseneck, Spreader hardware, Thru-Deck, and Mast Step. Except for the Mast Column, all other components are fabricated from remnants of the hull framing material. No costly manufactured hardware to purchase.

Each component will be broken down into Parts. For example, the Mast Head consist of the following parts:

- Reinforcing Plate
- Access Cover Flange
- Access Cover
- Tangs - Stays
- Tangs – Shrouds

'Bend Allowance' and 'Bend Deduction' formulas are used to develop the flat patterns, calculating the cut size of the part, locations of the bend line that form the part, and the location of interior features on the part.

Developing sheet-metal surfaces to a flat pattern is a subject onto itself and beyond the scope of this book. This subject, however, is covered in another of my books.

'Applied Metal Boatbuilding Methods – Sheetmetal Pattern Development'

Mast Column

The Mast Section shown below is 48-feet from Keel step to Mast Head. The below dimensioned drawing illustrates a conventional rectangular steel a mast section with radius corners. The Center Fabrication Spacer serves no calculated structural purpose. It is a fabrication aid only.

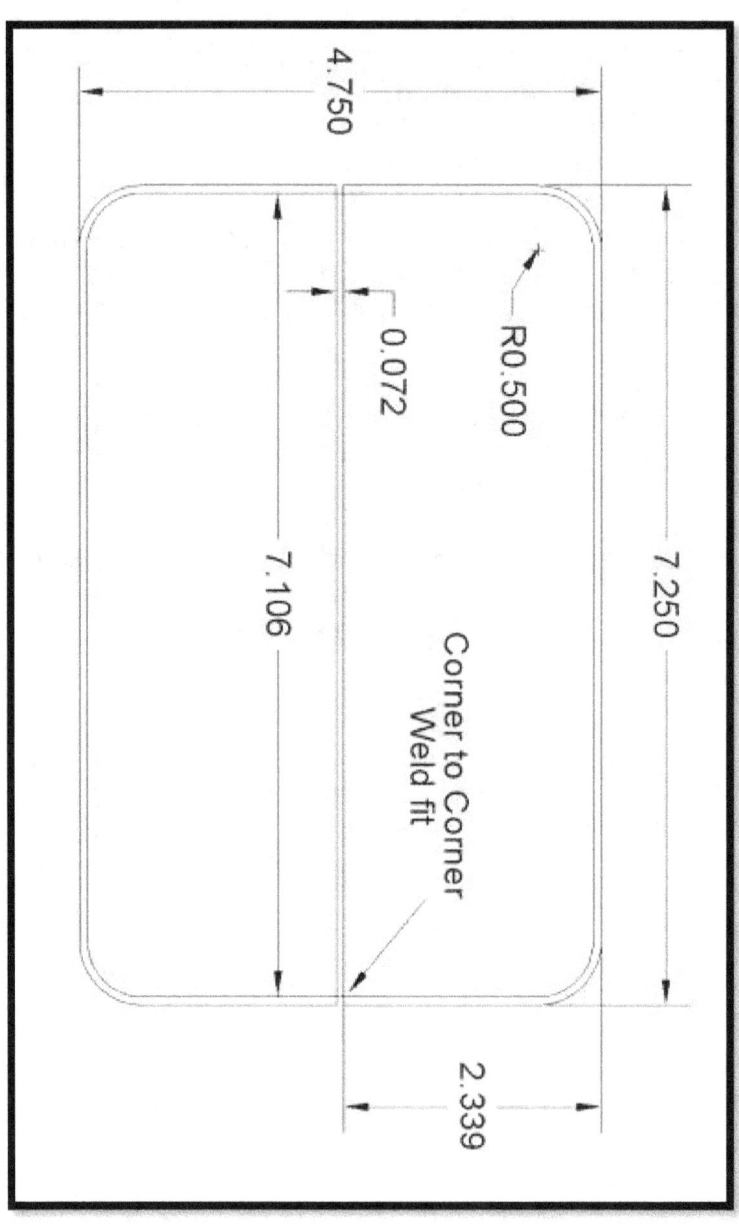

The Mast Columns vertical layout is shown in below.

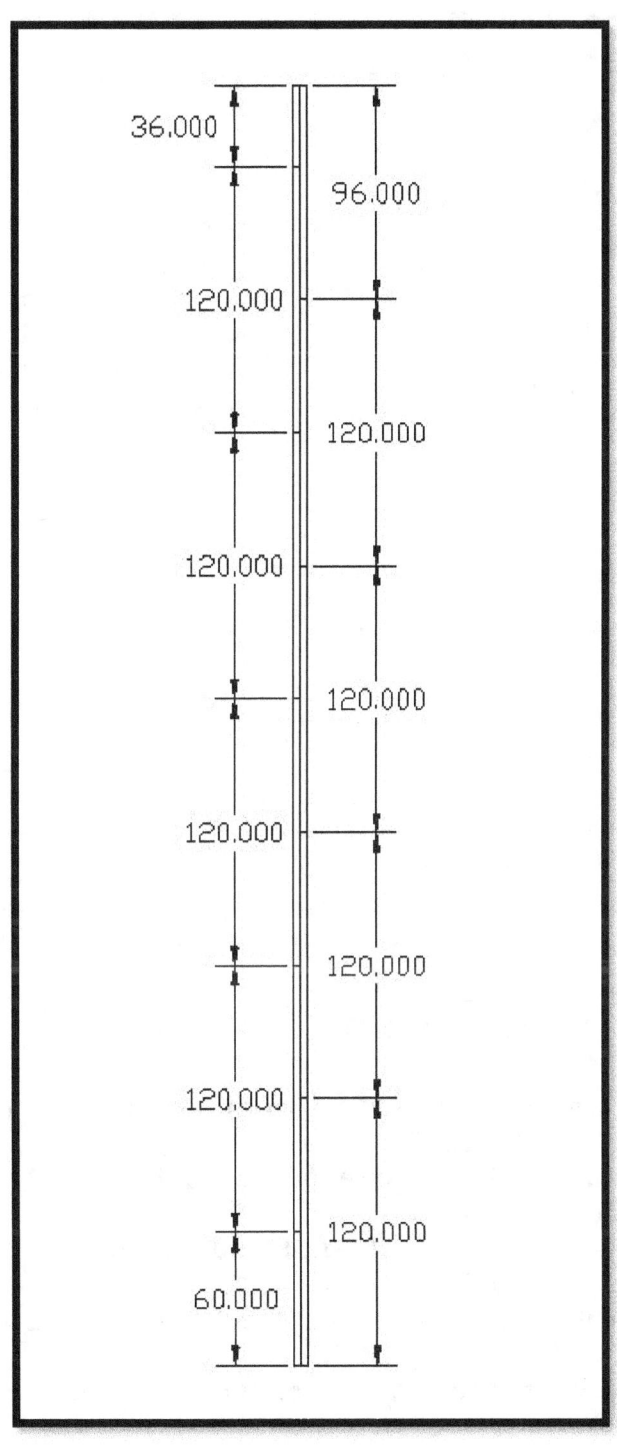

Mast Half Column - Flat Layout:

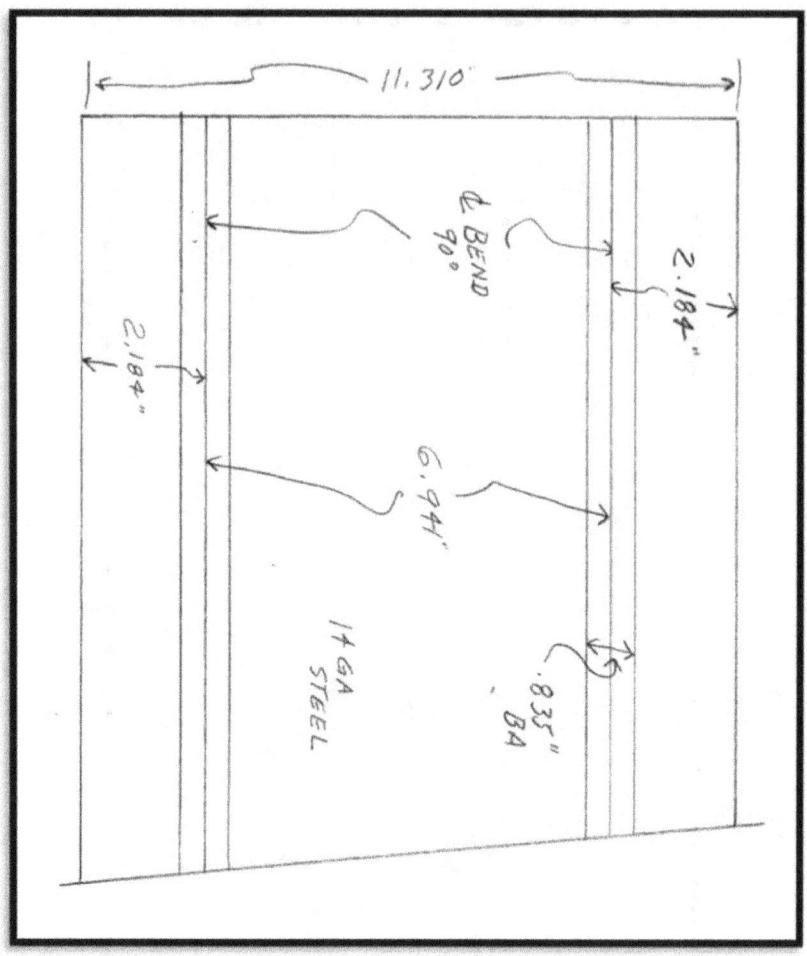

Cut List - Mast Column:

- **Half Section of Mast Head Column** - 8 piece 11.310" x 120" x 14ga.
- **Half Section of Mast Head Column** - 1 piece 11.310" x 96" x 14ga.
- **Half section of Mast Head Column** - 1 piece 11.310" x 60" x 14ga.
- **Half Section of Mast Head Column** - 1 piece 11.310" x 36" x 14ga.
- **Center Fabrication Spacer** - 5 pieces – 7.106" x 120" x 14ga.

Notes

Mast Head

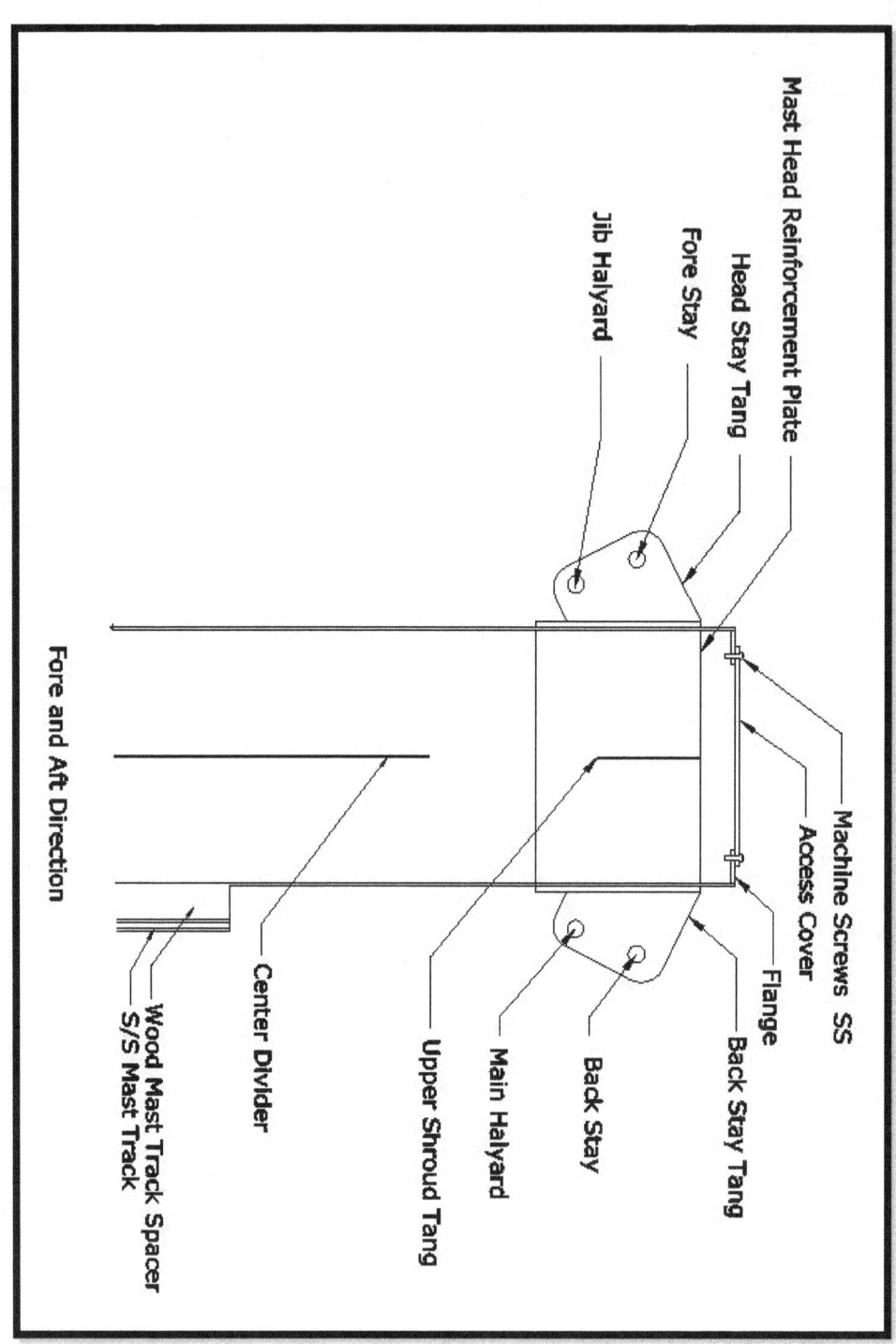

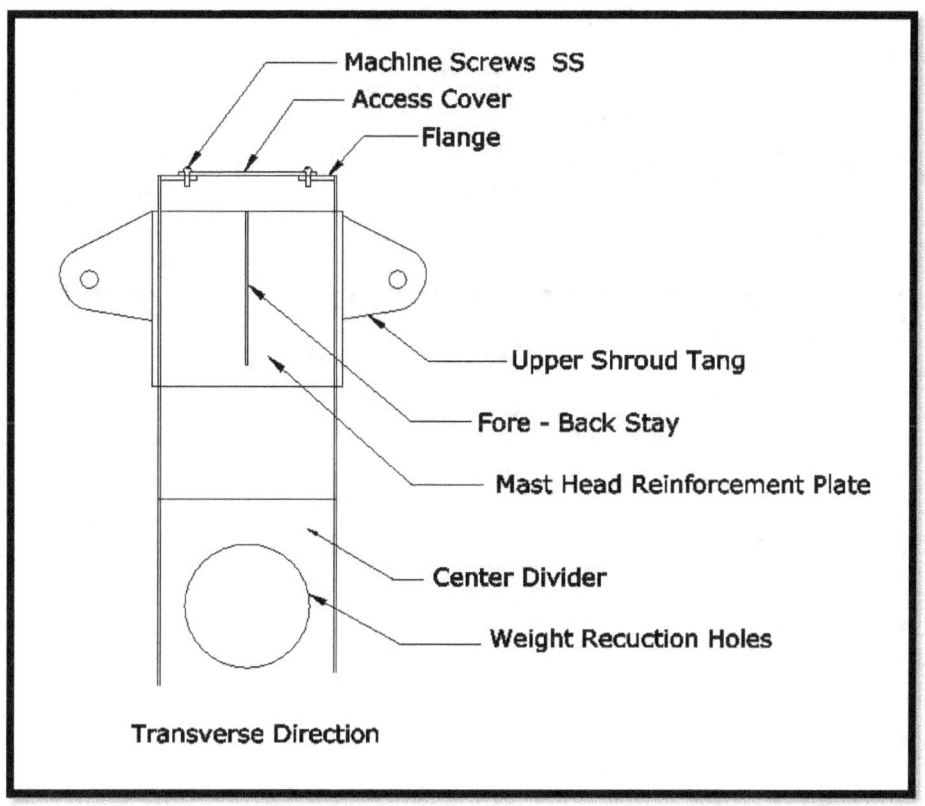

Mast Head Reinforcing Plate:

The Mast Head Reinforcement Collar is formed to the shape of the mast and needs to fit closely to the Mast Column. A calculated tight fit however would be a fabrication nightmare. Some clearance between the Mast Column and Reinforcing Plate is practicable from a fabrication Point of View.

Since, the Outside Radius of the Mast Column is 0.572", the theoretical Inside Radius of the Reinforcing Plate should be 0.572". This would be a dead on fit and again a fabrication nightmare. In addition, the diameter of the top punch to form a 0.572" radius would be 1.144". No metal fabrication shop is going to have a top punch of this diameter to form the Reinforcing plate.

Knowing that a 'Dead On' fit is impractical increasing the inside radius to 0.625" will give a little clearance between the mast and reinforcing plate by 0.053".

Choosing an inside radius of 0.625" for the Reinforcing Plate also resolves the tooling problem. Most metal fabrication shops will have a 1.250" round top punch. I find these changes to be acceptable.

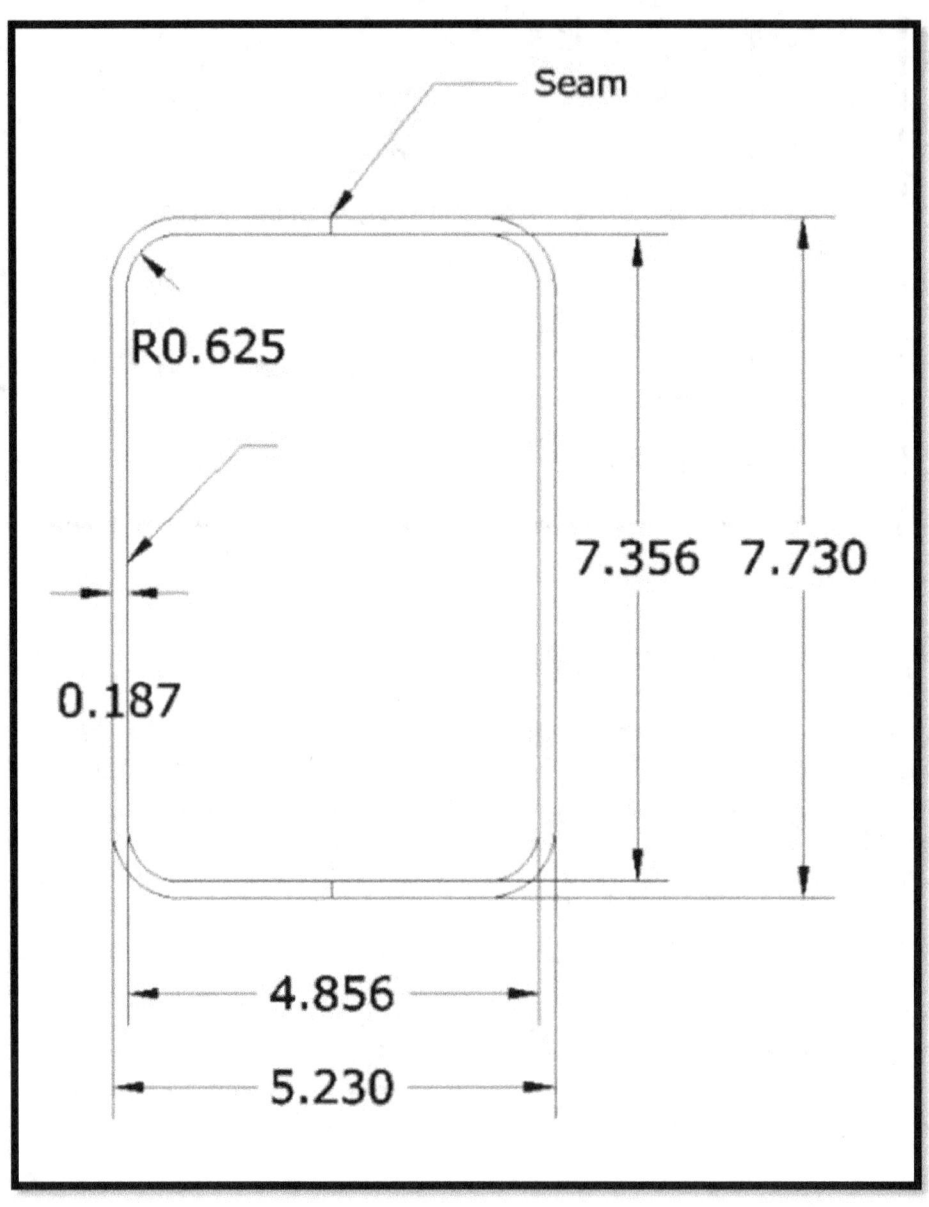

The developed flat layout for the Mast Head Reinforcing Plate is shown below.

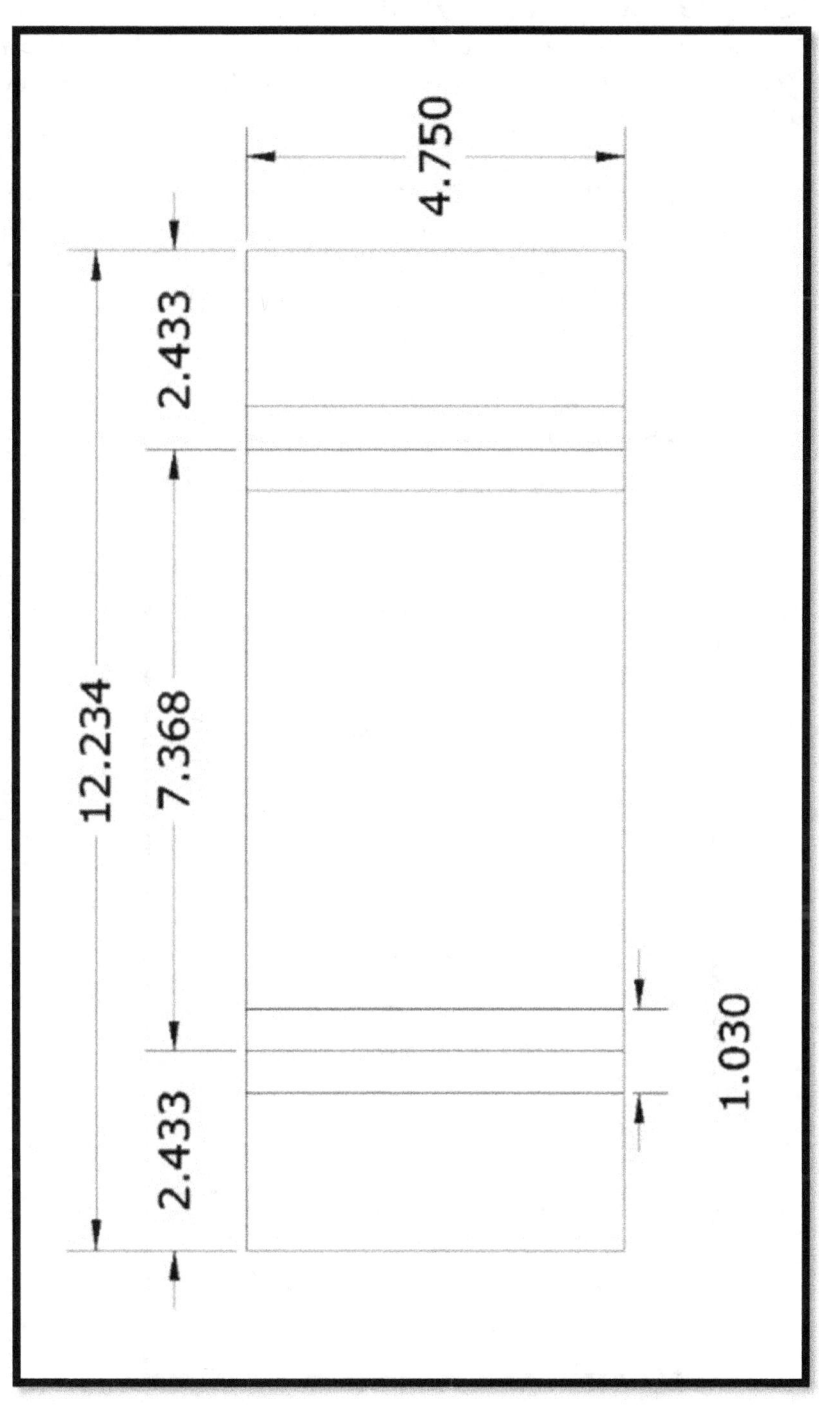

Access Cover & Flange:

Access to the interior of the mast is located at the top of the mast.

Mast Head Flange – Flat Layout: The mast head flange is thicker than the Mast Column. It is cut to the shape of the mast section and fully welded to the top of the mast column. The Flange is a minimum of 1" wide.

Access Cover – Flat Layout: An Access Cover is the same thickness as the flange and is fastened to the mast head flange with stainless steel machine screws. A gasket and or sealant is used between the Access Cover and the Mast Head Flange making this connection air and watertight.

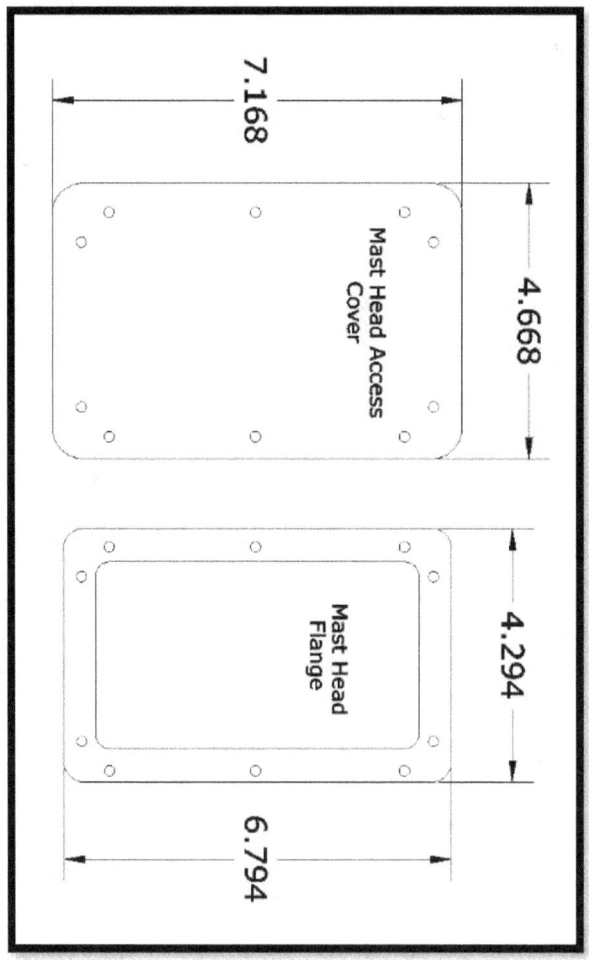

Tangs - Flat Layout:

The Upper Shrouds, Fore Stay, and Back Stay Tangs are fabricated of a thickness that will bear the compression on the mast. The upper holes attach the Stays. The lower holes attach the Main and Jib halyards or a Roller Reefing System. Below is the layout of the Mast Head Stay and Upper Shroud tangs.

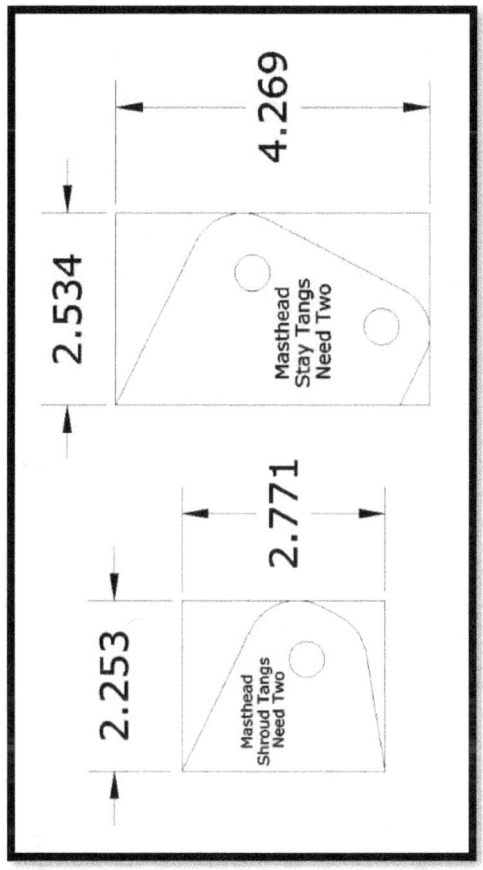

Cut List - Mast Head Components:

- **Reinforcing Plate:** - 2 pieces – 12.234" x 4.750"x 0.187"
- **Access Flange:** - 1 piece - 6.794 " x 4.294" x 0.125"
- **Access Cover:** - 1 piece - 7.168" x 4.668" x 0.125
- **Tang-Stays** - 2 piece - 4.269" x 2.533 x 0.250".
- **Tangs–Shrouds** – 2 pieces - 2.771" x 2.253" x 0.250"

Notes

Spreader's

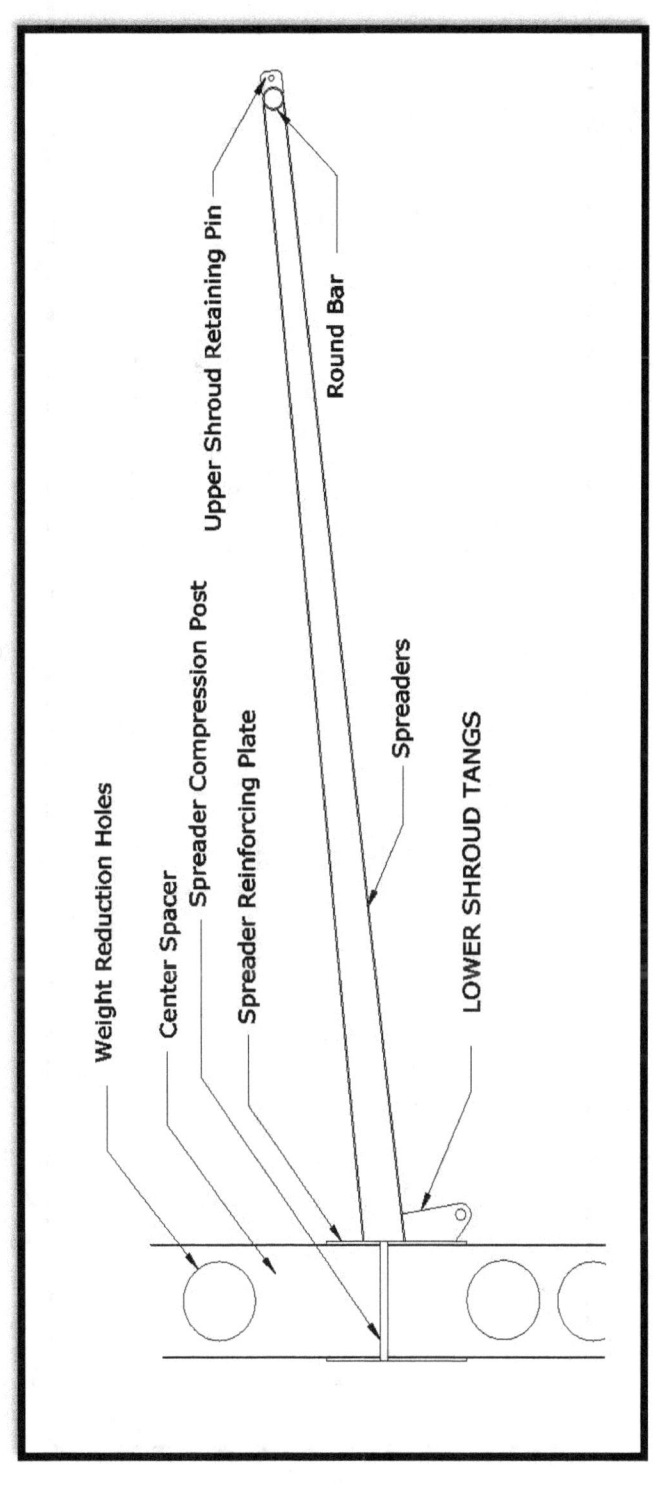

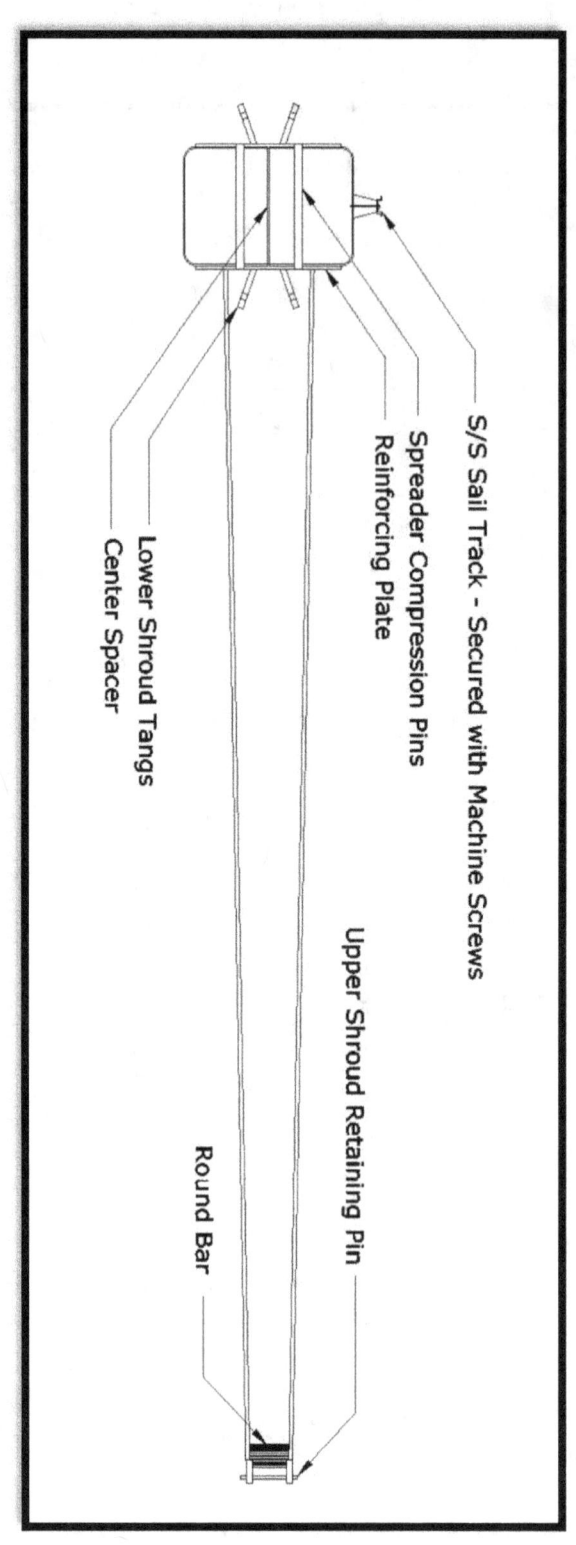

Spreader Compression Posts: Consist of 0.500" steel pins between the Spreader Reinforcement Plate.

Spreader Reinforcement Plate: The Spreader Reinforcement Plate is located on the side of the mast that is in contact with the Spreader and Shroud Tangs. The Spreader Reinforcement Plate is fully welded to the Mast Column.

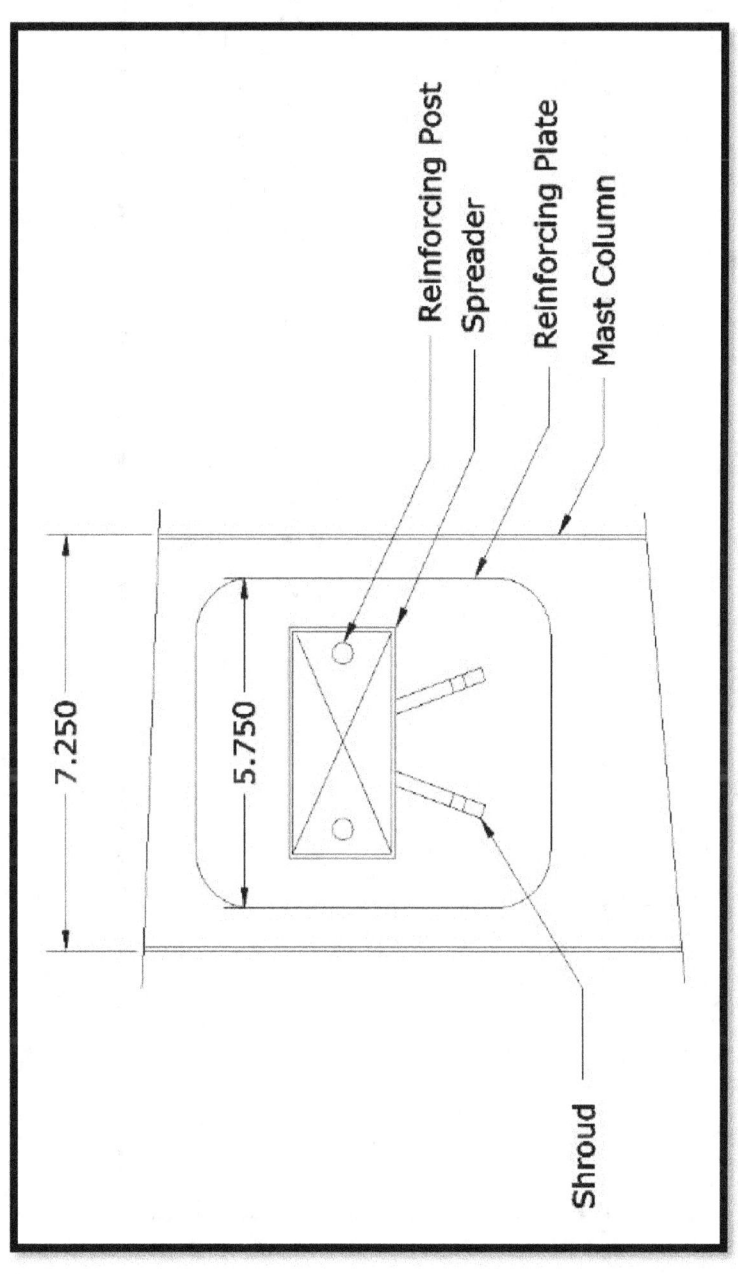

Spreader - Flat Layout: The Spreaders are fabricated from steel sheet material the same type and thickness as the Mast Column according to the Architectural Drawings. The Spreaders are fully welded to the Spreader Reinforcement Plate.

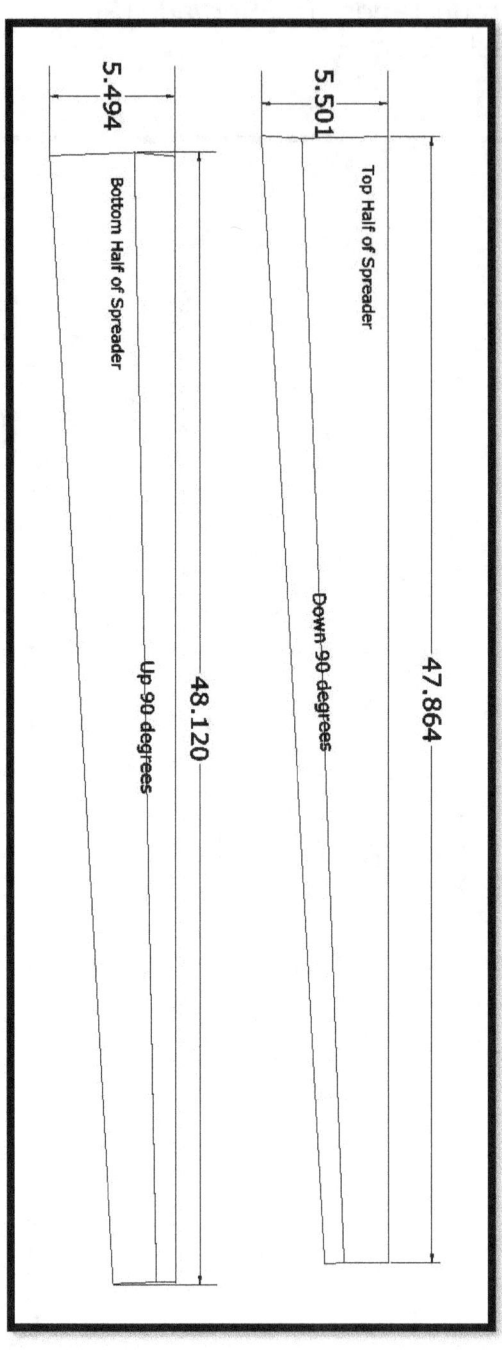

Lower Shroud Tangs - Flat Layout:

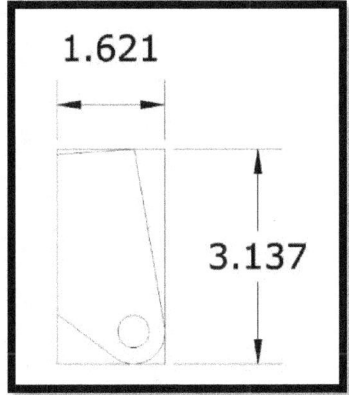

Spreaders Ends: A Round Bar is weld at the Spreaders ends for a soft landing for the 1 x 19 Stainless Steel Standing Rigging. A pin at the Spreader ends secures the Standing Rigging.

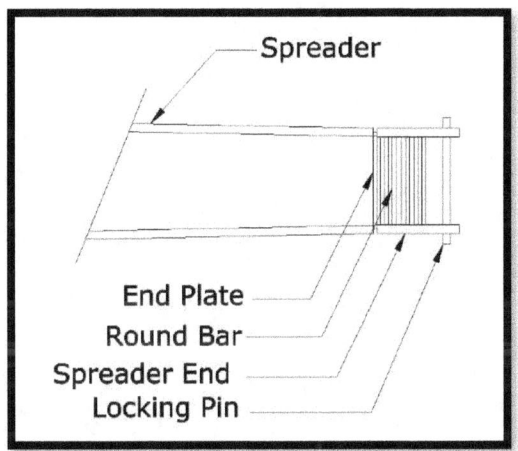

Cut List – Spreaders:

- **Reinforcing Plate:** - 2 pieces – 5.750" x 6.500"
- **Reinforcing Post:** - 2 piece – Width of Mast
- **Spreaders - Top:** - 2 piece - 47.864" x 5.501"
- **Spreaders - Bottom** - 2 piece – 48.120" x 5.494"
- **Lower Shrouds Tangs** – 4 pieces – 3.137" x 1.621"

Notes

Boom Gooseneck

Boom Gooseneck: A three-dimensional pictorial of the Boom Gooseneck is shown below. It is fabricated from 11-gauge Stainless Steel and Pipe or Tubing that will accept a 0.500" pin.

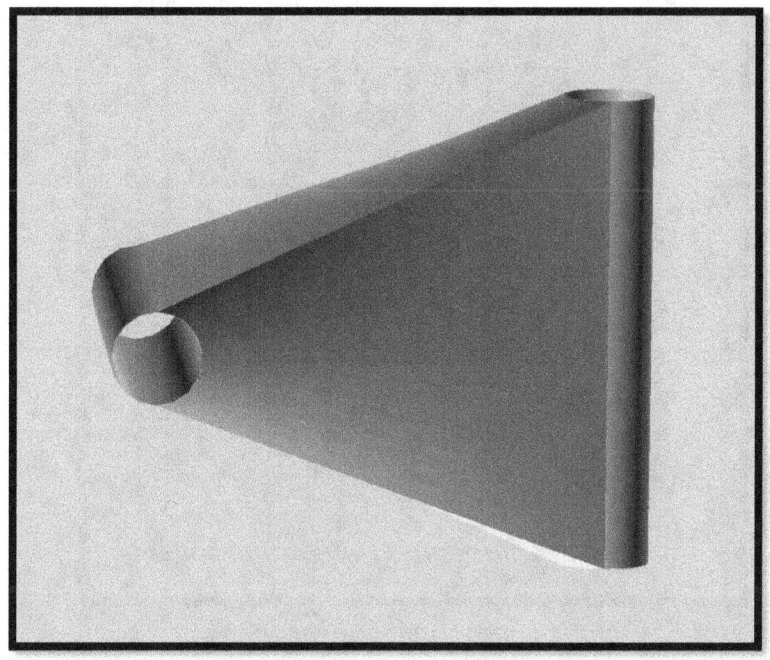

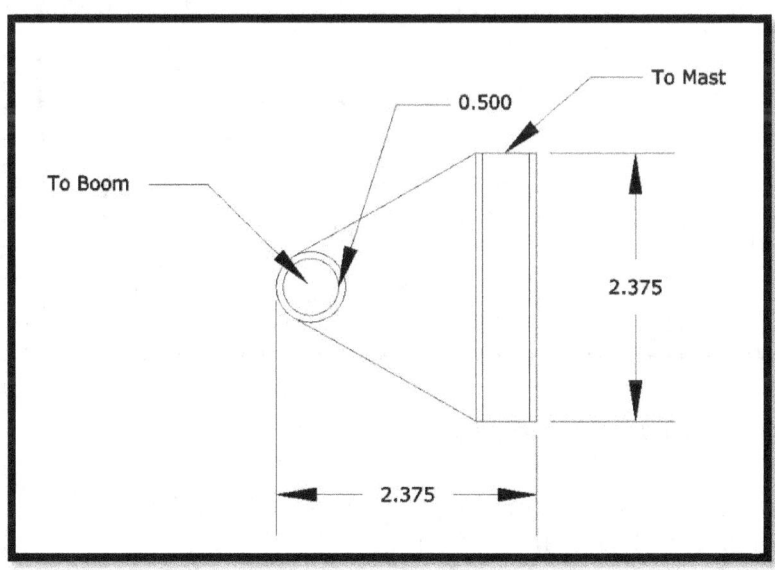

Reinforcing Plate: A 0.187" thick is welded to the mast column
in way of the Boom Gooseneck Brackets.

Boom Gooseneck Support Brackets: Are fabricated from 0.187" thick steel. The full-size layout is shown below.

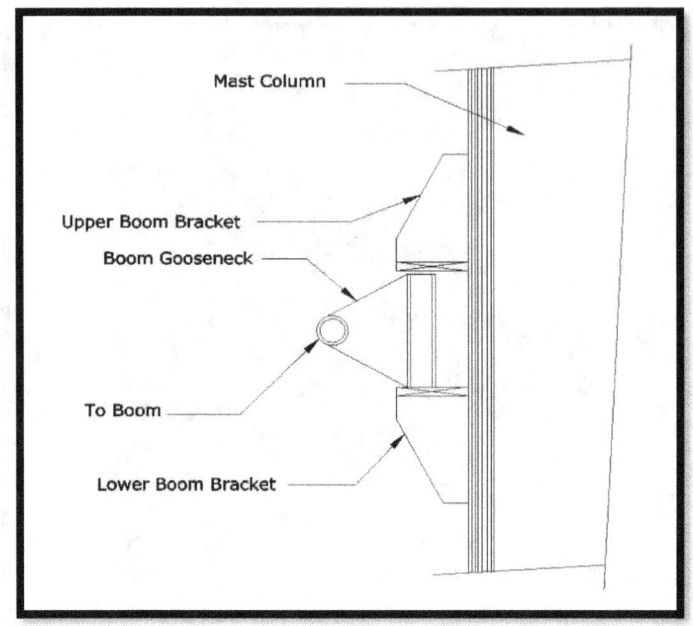

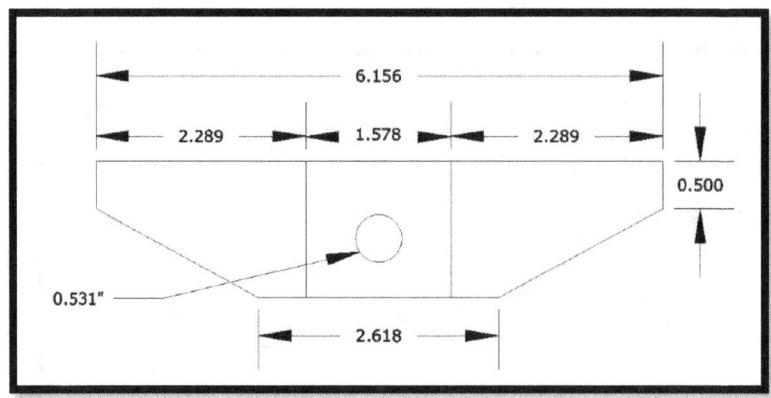

Cut List – Boom Gooseneck:

- **Reinforcing Plate:** - 1 pieces – Cut to Suite
- **Support Brackets:** - 2 piece – 6.156" x 1.437"
- **Boom Gooseneck:** - 4 piece - Cut to Suite
- **Boom to Mast Tubing** - 2 piece – 3.375" Long

Notes

Thru Deck

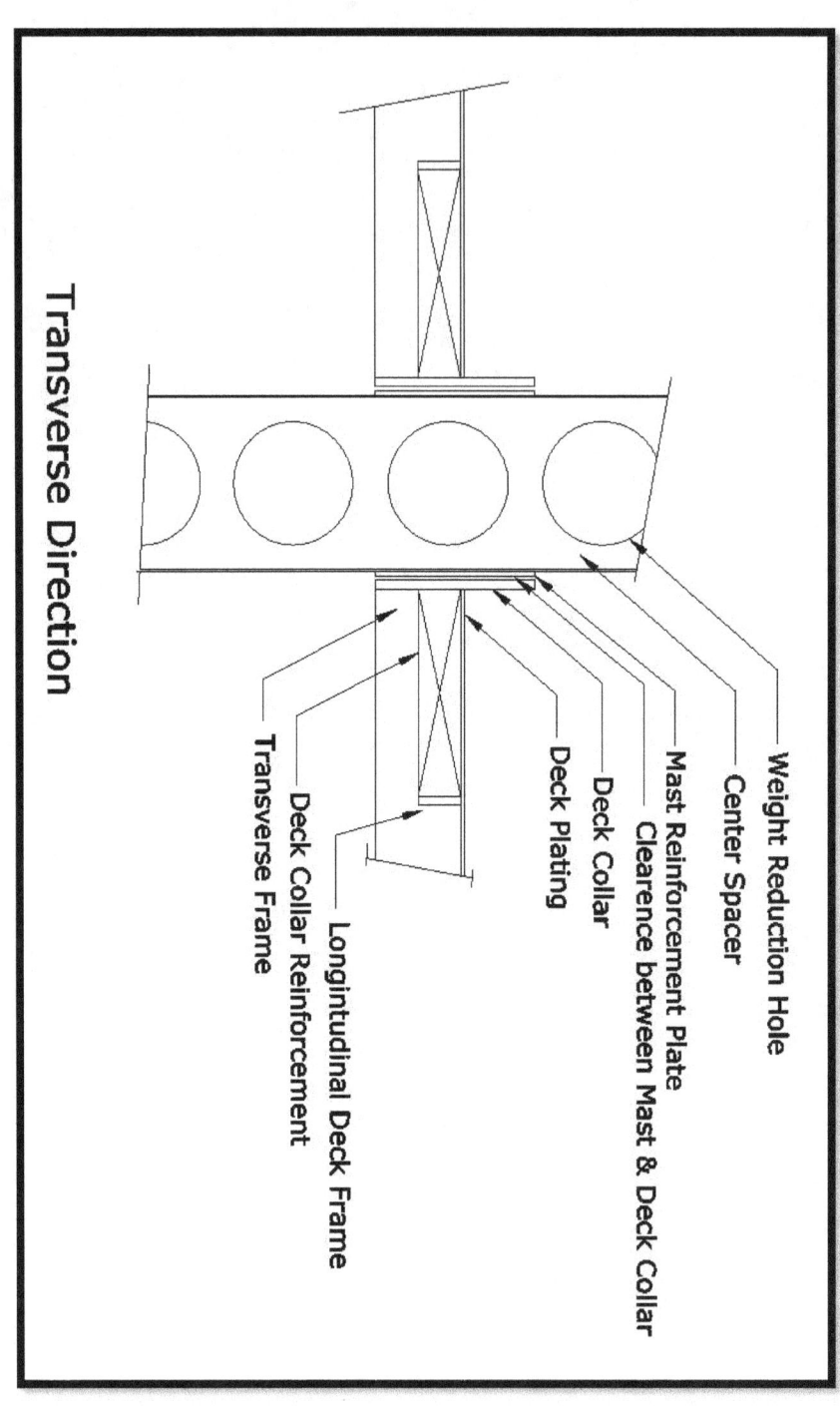

Thru-Deck Mast Reinforcing Plate: A Reinforcing Plate is located where the mast penetrates the cabin trunk top. It is dimensioned the same as the mast Head reinforcing plate.

Mast Clearance: A reasonable amount of clearance, 0.312" shown in this drawing, is needed between the Mast and the Thru- Deck Collar for the installation or removal of the mast.

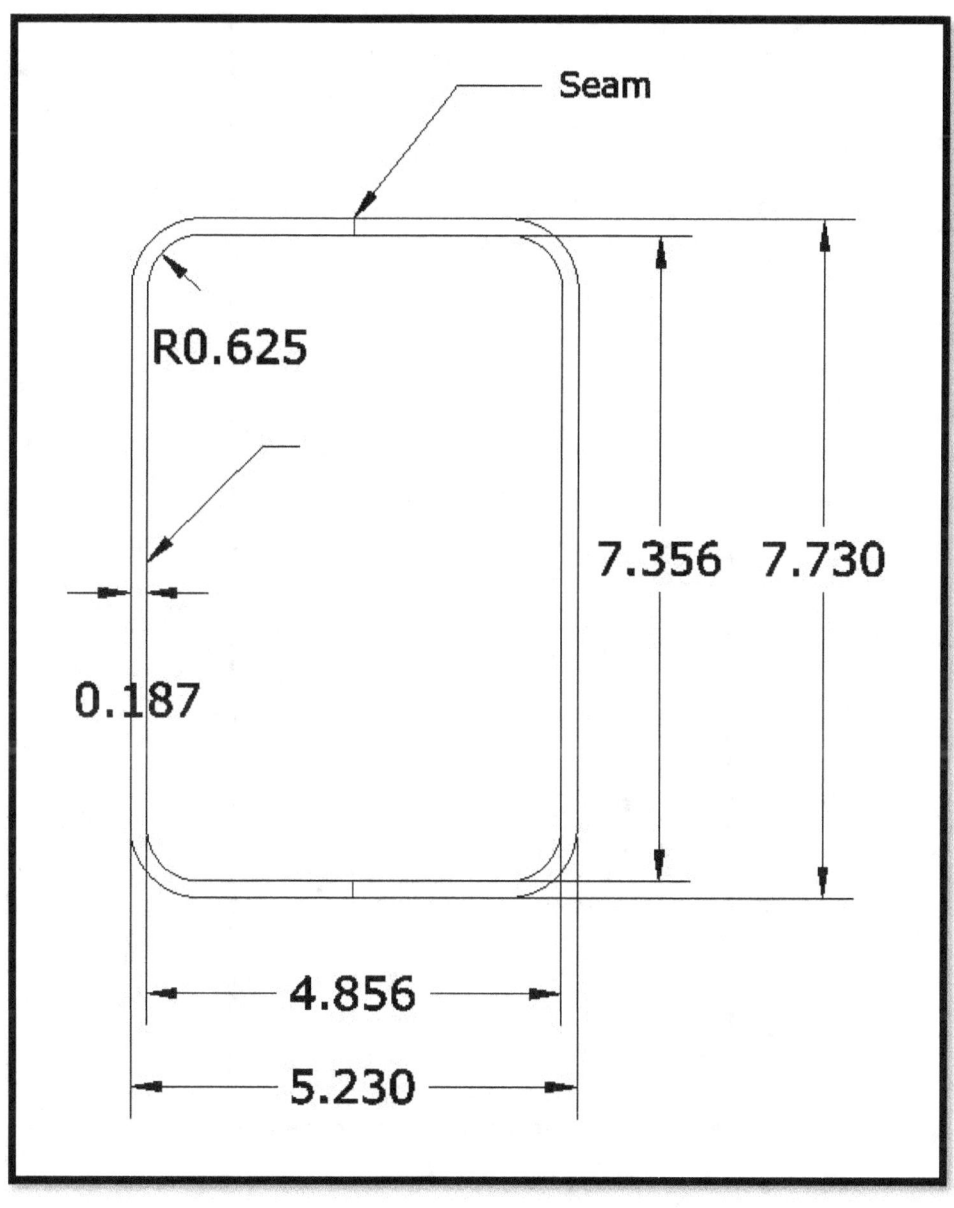

Thru-Deck – Collar Layout: The Thru-Deck Collar extends at least 2.000" above the deck surface. Inside, it extends to a depth equal to the width of the framework. Note: This component is *not* part of the mast, but part of the cabin structure. The outside dimension of the deck collar is: 12.412" x 8.834". The inside radius is 1.375" with a material thickness on 0.250". The flat layout is shown below.

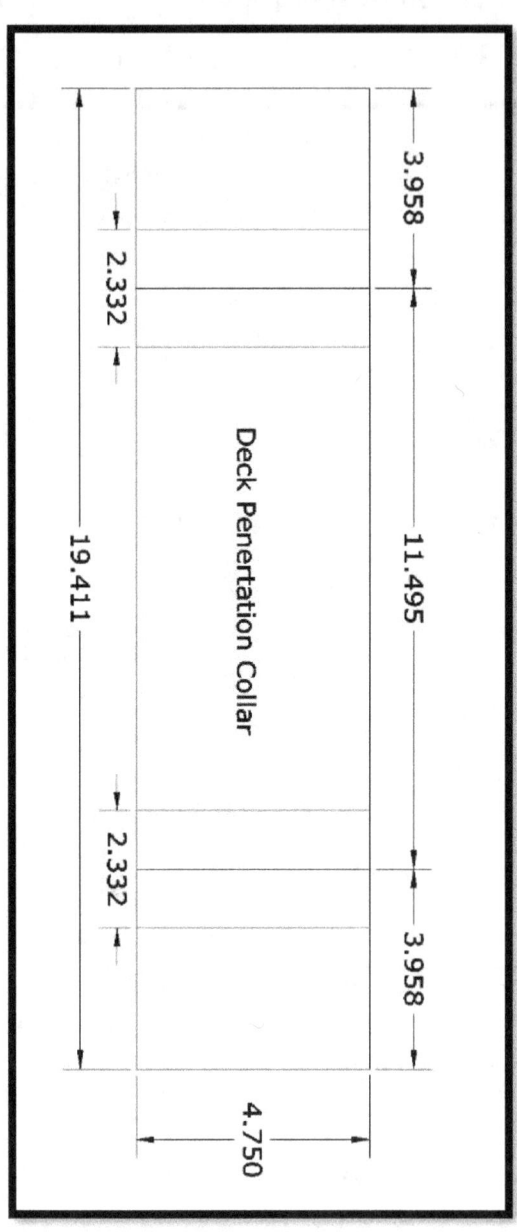

Deck Structure: The Mast Collar cannot be supported by the Shell Plating alone. In an ideal world the mast will lie in way of a transverse frame, and be supported by that frame. The drawing below illustrates additional ways to support the Mast Collar where the mast collar penetrates the deck plating.

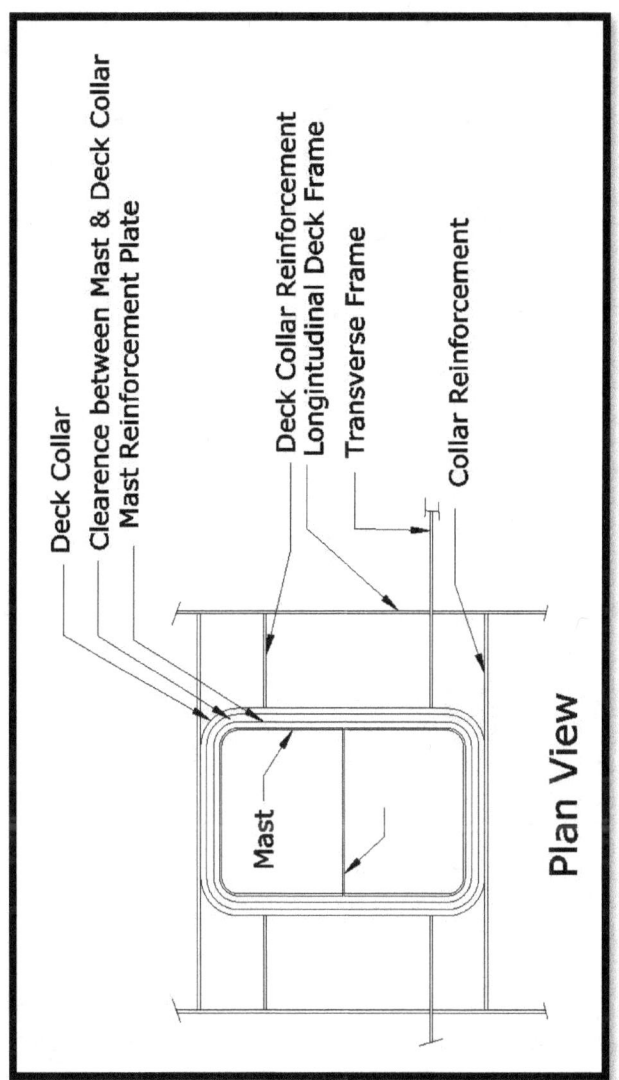

Cut List – Mast Penetration:

- **Reinforcing Plate:** - 2 pieces – 12.234" x 4.750"
- **Deck Penetration:** - 2 piece – 19.411 x 4.750"

Notes

Mast Step

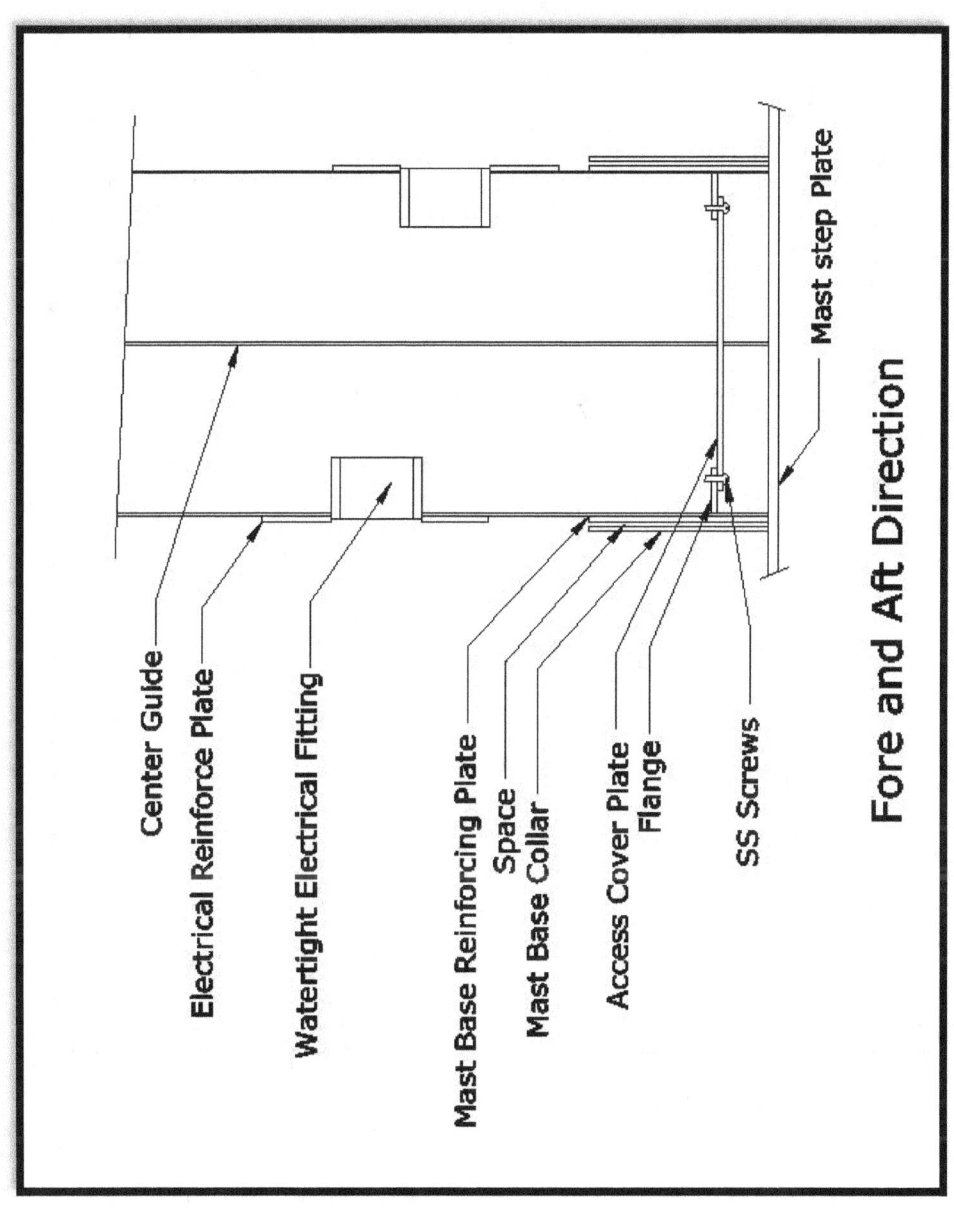

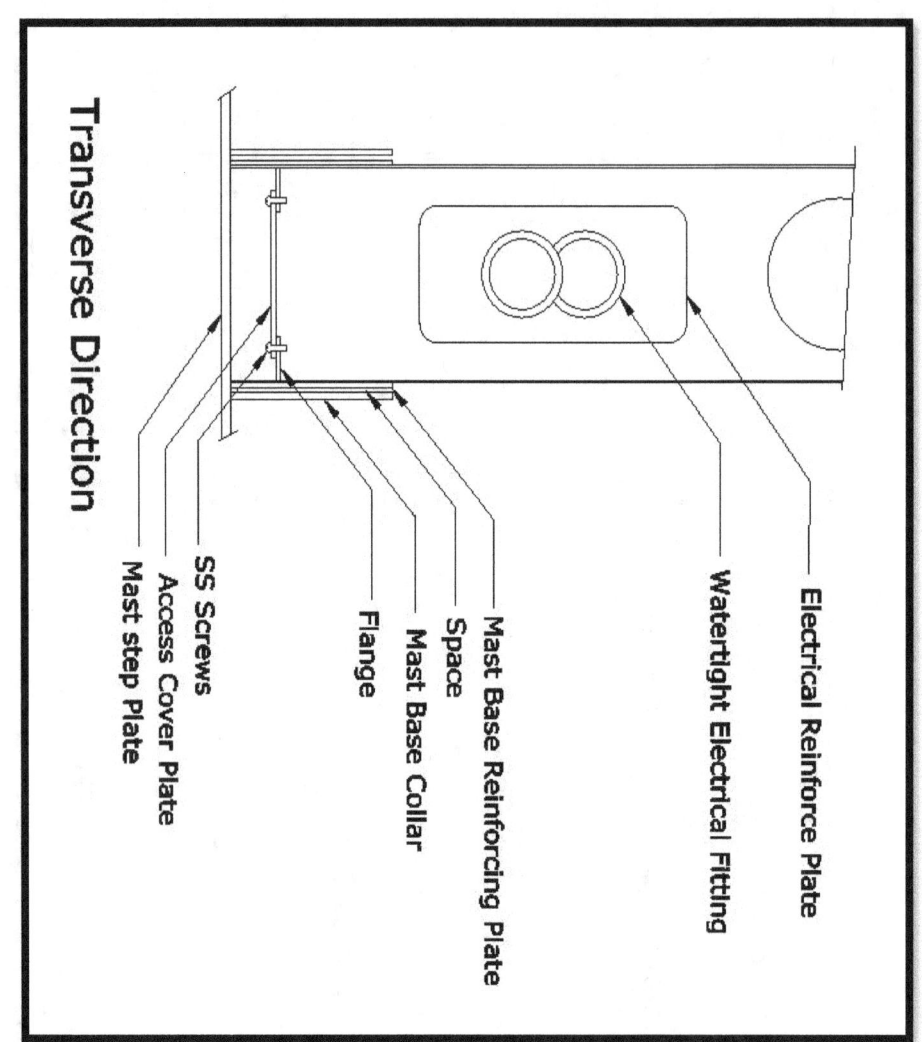

Mast Step Reinforcing Plate: The Reinforcing Plate at the mast step is dimensioned the same as the mast Head reinforcing plate.

Mast Clearance: A reasonable amount of clearance between the Mast and the Mast Step Collar. The Clearance here does not need to be as large as the clearance at the Deck Penetration.

Mast Step Collar: The Mast Step Collar will sit on the Keel Step structure, not shown. The height of the Mast Step Collar is a minimum of 2.000".

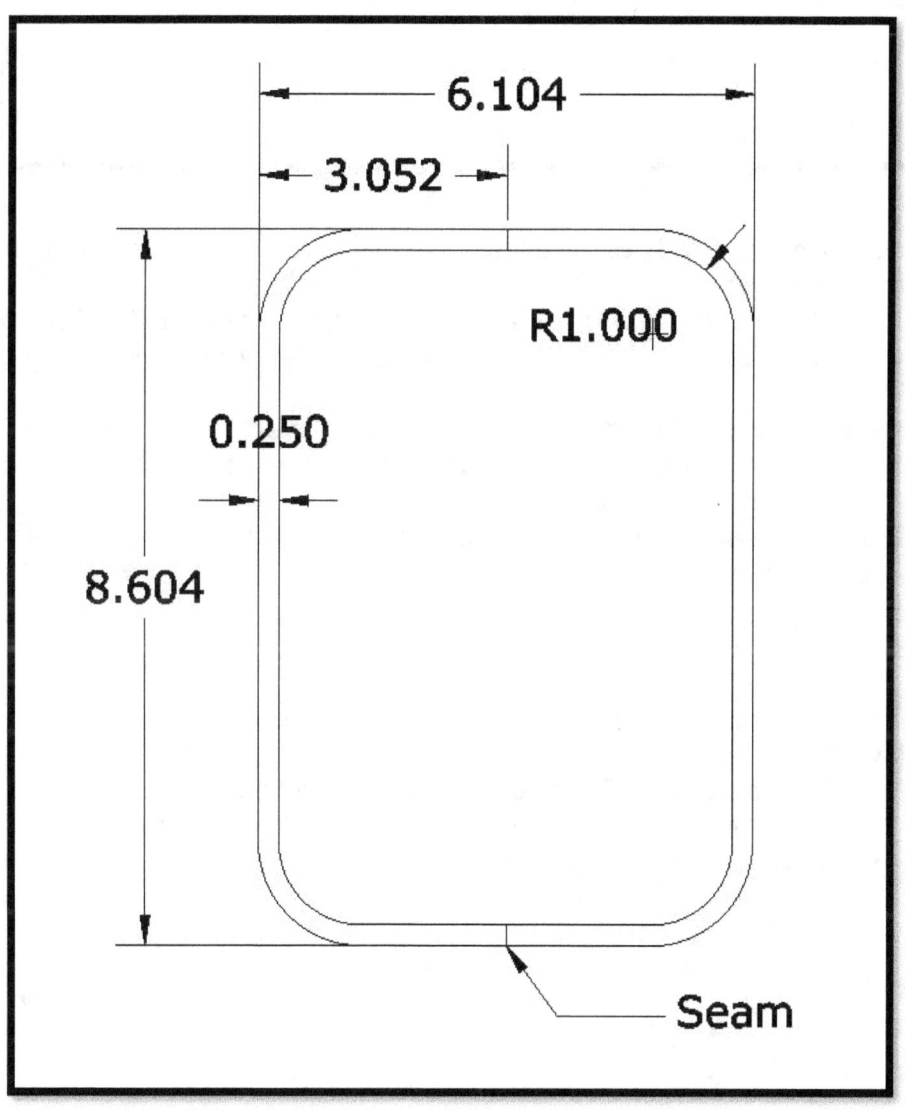

Mast Step Collar - Flat Layout:

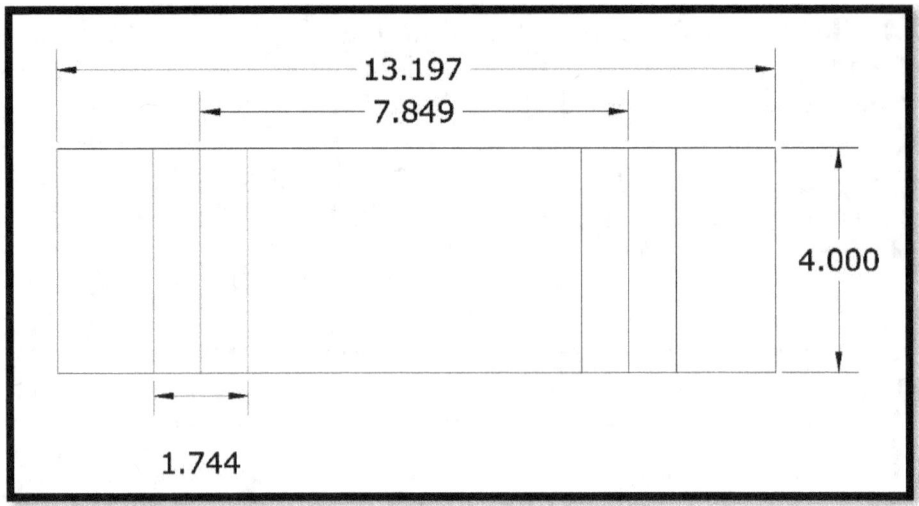

Access Cover and Base Flange: Mirrors the Access Cover and Flange at the mast head. The difference is that the flange and Access Cover is inside the mast column by a minimum distance of 1.000" from the mast step plate. The Access Cover needs to be water and air tight.

Electrical Exit Fittings: A reinforcement plate is place where the electrical wiring exit the mast column. The drawing shows these fitting at the base of the mast. Another location is between the cabin deck shell plating and the interior headlining inside the cabin. These fitting will be made air and water tight.

Cut List – Mast Step:

- **Reinforcing Plate:** - Same as Mast Head
- **Mast Step Collar:** - 2 pieces – 13.197" x 4.000"
- **Access Cover Flange:** - 1 piece – Inside of Mast Column
- **Access Cover:** - 1 piece – Inside of Mast Column
- **Electrical Fittings:** - 2 pieces

Notes

Standing Rigging

Typically, the Standing Rigging is '1 x 19' Stainless Steel Cable. Both ends of the cable will have Swagged Fitting or Sta-Lok fittings. In the arrangement shown, both the mast head fitting and the chainplate fitting have 'Eye' ends.

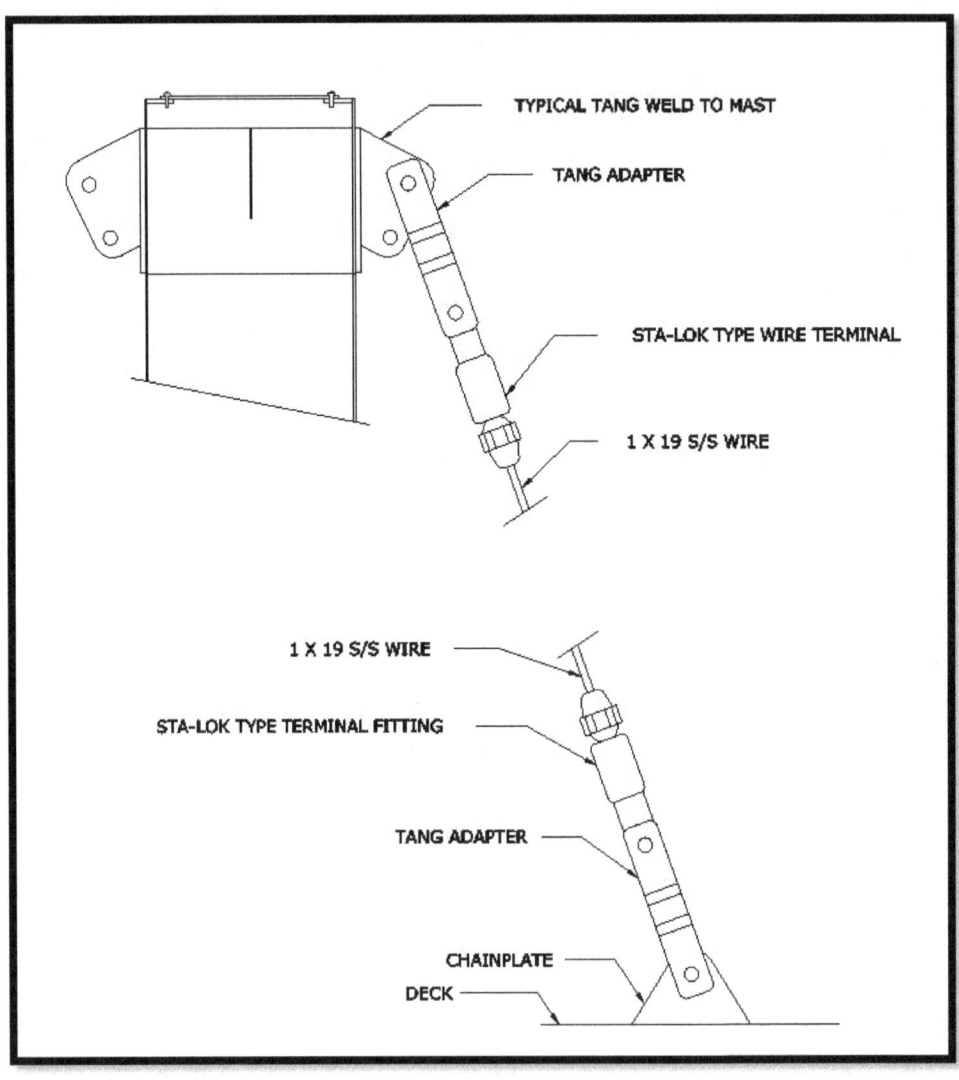

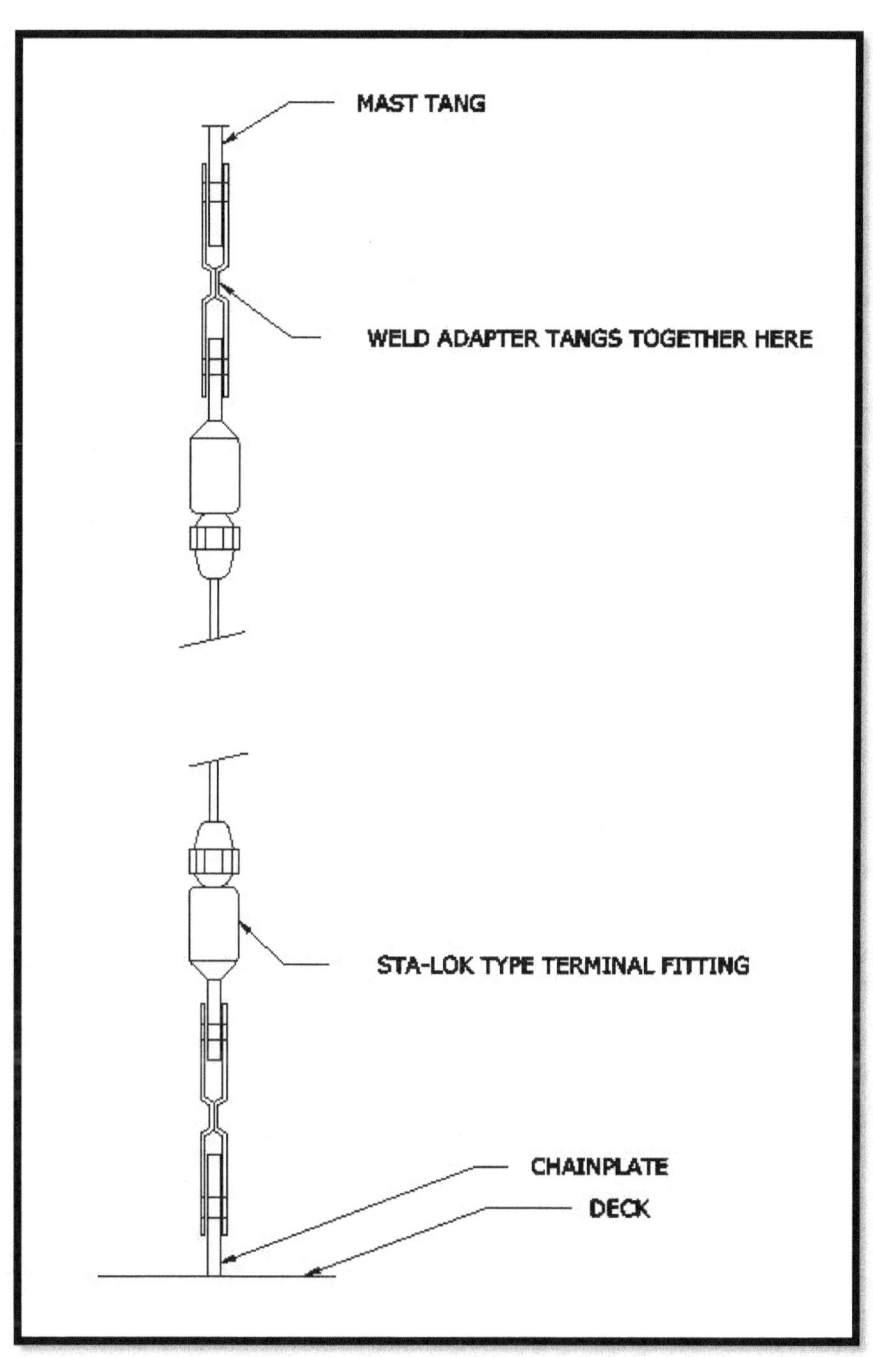

Tang Adapters: Stainless Steel Tang Adapters are shown between the end terminals of the standing rigging terminals at both the mast stays, shrouds, and Chainplates.

Notes

Fabrication of the Mast Column

The component involved in fabricating the Mast Column are:

The **Starting half section** is five feet long. It begins the mast construction. Its purpose is to stagger the mast half sections to guarantee that no two seams fall across from the other.

The **Full sections** are of 10-feet in length, comprising the bulk of the half sections used to fabricate the mast column.

The **Last two sections** at the top of the mast vary in length to achieve the desired mast height.

The **Transverse Divider** is an aid in fabrication giving the half sections shape during fabrication. It serves no calculated structural purpose.

The corners of the mast are formed to a 90-degree angle. In practice, a bend of 1 or 2 degrees over 90 degrees is more workable when fabricating the mast. Never under-form components. It is always easier to take bend out of a part than to put bend in a part during fabrication.

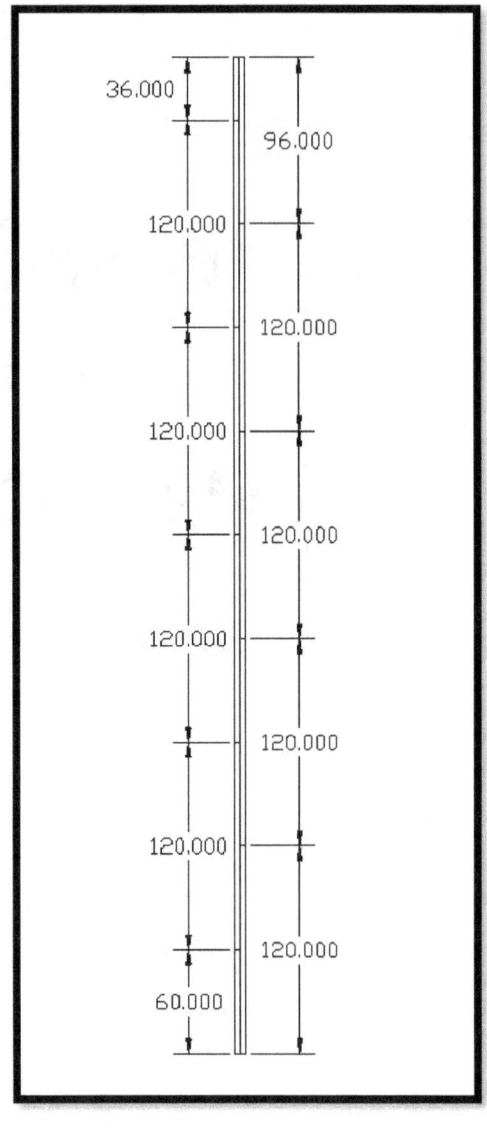

Since the quality of the mast starts at the beginning of the fabrication, the sheet material must initially be cut squarely, dimensional correct, and formed to a consistence angle along the entire length.

> My preference of welding electrodes is type 6011. This electrode provides excellent weld penetration into the base metal. It also does not lay on a heavy weld bead.

Mast Construction Procedure:

Referring to the below drawing. Start the fabrication with a 10-foot half section and tack weld a 7' 6" transverse center divider to it. When tacking these two pieces together, be sure there are no twists or other distortions. Any twist in this first section will carry through the entire length of the mast.

- To the above, tack weld the 5-foot half column section.
- To the 10-foot section add a 10-foot center fabrication section.
- You should now have a fabrication that looks like the drawing below.

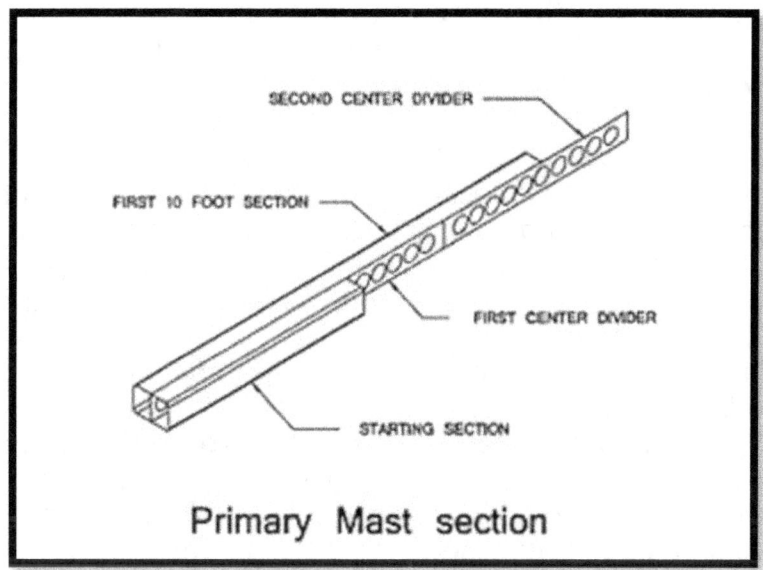

Primary Mast section

- Before proceeding, check that this initial fabrication is free of twists by carefully sighting down its length with your eye. Your eye is the discriminating instrument that maintains a straight and twist free finished mast.
- Follow this procedure of staggering fabrication center dividers and mast sections until the designed mast height is reached. Note: No seams will ever align with each other.
- Add the Reinforcing plates at the masthead, spreaders, boom location, deck penetration, and keel step.
- Add all other components to the reinforcing plates.

Tack Welding:

- Tack welds should always be small, perhaps no longer than 3/16".
- All components and the mast column are to be fully tacked welded before continuous welding of the mast commences.
- Fully tack welded means that the space between tack welds will be no more than 1.250" apart and closer where appropriate.

Staggered Welding Sequence:

A '**Staggered Welding Technique'** is use to Continuously weld all sections of the mast fabrication. *Do not deviate from this process.*

- Staggered welds shall be no longer than 2.000".
- To start, choose a location anywhere along the length of the mast, placing your first 2.000" weld. Skip a space of at least 12.000" from the first weld and make another 2.000" weld. Skip again, and make the last of a three-weld sequence.
- Move to the opposite side of the mast column and repeat the three-weld sequence opposite the first sequence.
- The next sequence is a distance from the first. For example, if you started the welding of the mast around the center, move near the top or bottom of the mast column and repeat the three-weld sequence there.
- Move to another location on the mast column, maybe near the bottom of the mast, and repeat the three-weld sequence there.
- Keep repeating the three-weld sequence until the entire mast column has been welded.
- Follow the same procedure when continuous welding the mast components.

Finishing the Mast Surface:

Mast integrity is vital, **do not** grind any welds flush to the surface of the mast, especially along the length of the mast column itself. Removing weld splatter and smoothing of the welds is acceptable.

Notes

Bezier
True Round Designs

Bezier 28

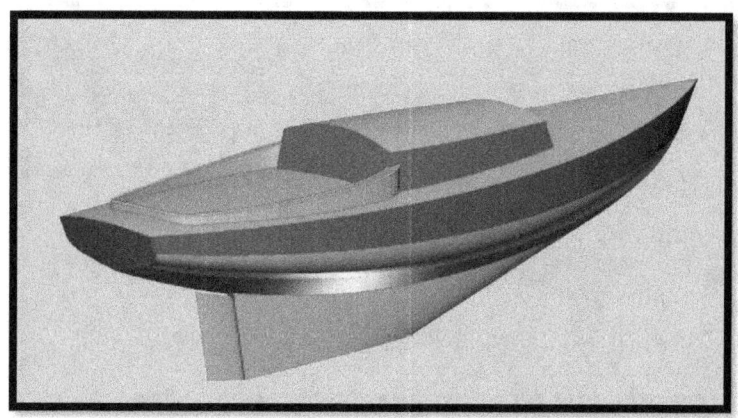

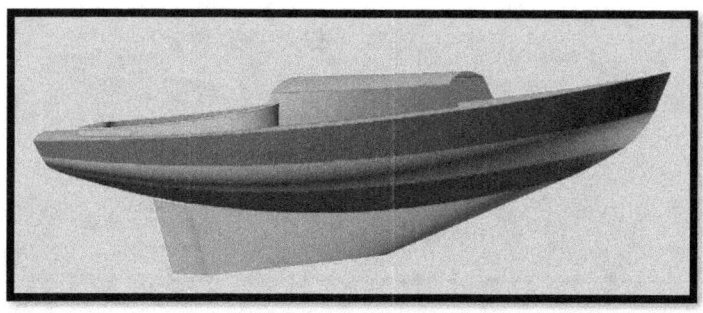

Bezier 35

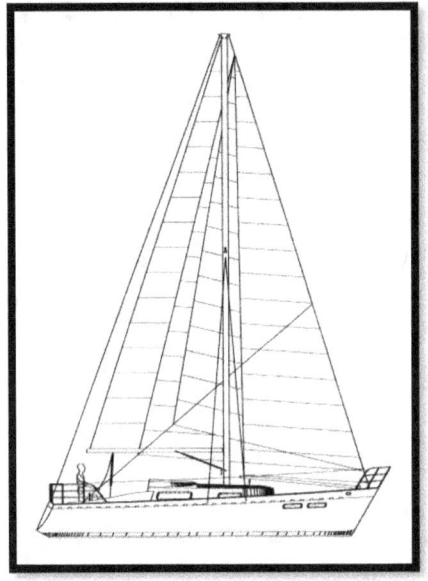

Bezier 12.5

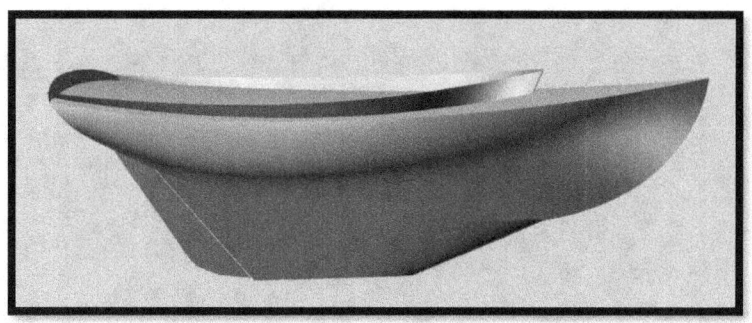

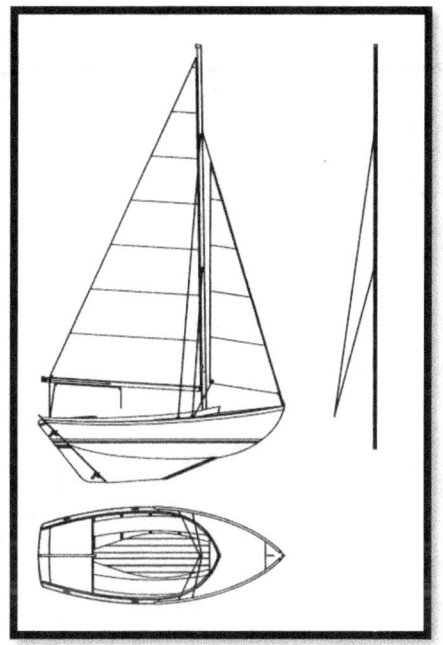

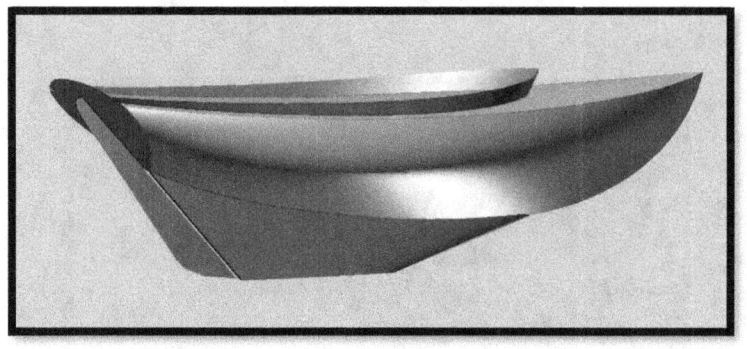

Bezier 34

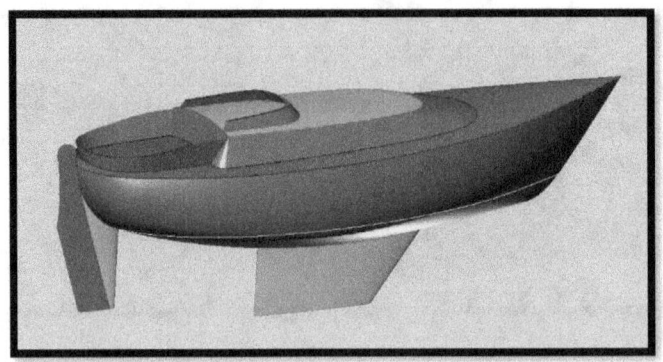

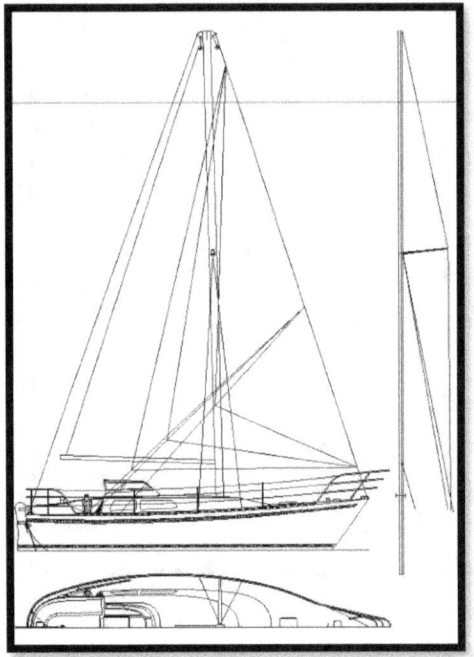

Bezier Books

Hard Copies at Amazon

PDF Versions at metalsailboats.com

True Round Metal Boat

Construction True Round

Metal Boat Design Applied

Metal Boatbuilding Methods

Converting Hard Chine Sailboats to True Round

Notes

Notes

www.ingramcontent.com/pod-product-compliance
Lightning Source LLC
Chambersburg PA
CBHW082119220526
45472CB00009B/2242